Crystals and Crystal Growing For Children

A guide and introduction to the science of crystallography and mineralogy for kids

Samuel Grundy-Tenn

Samuel Grundy-Tenn

Copyright © 2012 Samuel Grundy-Tenn

All rights reserved.

Although the author and publisher have made every effort to ensure that the information presented in this book was correct at the present time, the author and publisher do not assume and hereby disclaim any liability to any party for any loss, damage, or disruption caused by errors or omissions, whether such errors or omissions result from negligence, accident, or any other cause.

ISBN-13: 978-1539104407

CONTENTS

An Introduction To Crystals 8

    What Are Crystals? 8

    How Are Crystals Formed? 10

    What Are Crystals Used For? 12

Crystal Projects 14

    Salt Crystal Coral 14

    Epsom Salt Frost Crystals 17

    Growing Alum Gemstones.................................19

    Rock Candy Sugar Crystals................................21

    Table Salt Crystals...23

    Blue Copper Sulfate Crystals 25

    Blue-Green Copper Acetate Crystals 27

    Big Borax Colored Crystals.................................29

    Aragonite Crystals 32

    Color Changing Crystals.....................................34

    Bismuth Metal Crystals.......................................36

Sodium Nitrate Crystals 39

Further Projects 42

Final Thoughts 43

Crystals and Crystal Growing For Children

AN INTRODUCTION TO CRYSTALS

What Are Crystals?

What do you think of when you think of crystals? You might think of rich jewels or you may think of a hotel chandelier? Well you would be surprised at how many other things in the world are classed as crystals. Certain substances such as salt, sugar and snow are all classed as crystals.

The definition of a crystal is based from the scientific idea of atoms. This is the idea that everything in the world is made up of tiny little pieces called atoms, these atoms are far too small to see with the naked eye. Each material in the world is different from one another due to the way the atoms are arranged. In a crystal the atoms are arranged in a specific way, this means that not all solids are crystals.

A crystal is known as a substance with a repeated structure. This means that there is a pattern in the atoms that is repeated many times in order to produce the substance that you will see with your eye. What a crystal is made of depends on what type of crystal it is, snowflakes, salt and sugar are all crystals and all made from different materials. Snowflakes are made of water that has frozen, whereas salt is made from chemical elements called sodium and chlorine.

How Are Crystals Formed?

Crystals are formed in a number of ways, most crystals are made through evaporation such as in the process of salt water evaporating, evaporation is the process of when a liquid is heated and it is taken up into the air. Another way that crystals are formed is that when the liquid has been evaporated, the liquid will rise and under certain conditions the evaporated liquid, which is called vapor, will freeze and form crystals in the air. In the case of water the water vapor will freeze in the air and return the earth in the form of snowflakes which, as previously mentioned, are crystals. Ice can also form as frost on surfaces when the temperature is cold and the humidity is high, humidity is a measure of how much water vapor in the air there is.

A lot of the crystals that would spring to most people's minds are not made using any of the above methods, imagine weather of snowing diamonds, rubies and sapphires. Instead there crystals are made from melted rock deep inside the earth, when the melted rock cools down and becomes solid it sometimes will become a crystal. This occurs if when the rock is cooling bubbles are produced, the inside of these bubbles will grow crystals inside them.

In fact we have now created ways to grow the more

precious crystals that are made deep within the earth, we can artificially produce diamonds sapphires and rubies. This is done using two methods, one of the methods to produce an environment similar to that where each crystal is naturally made in and place the raw materials that the crystal is made of. This over the course of around 8 months will grow a crystal which is almost identical to that would be found within the earth. This is called Flux production. The other method is to take the raw materials drop them in a tube and melt them at extremely high temperatures, the chemicals will fuse together to form the desired crystal. This method of artificial crystal production is called flame fusion.

    As a crystal is grown the pattern that it follows is repeated this means that it will keep the same shape as it grows larger, as I said before the chemicals that the crystal is made of determine which crystal is formed when it grows. When crystals start growing they may look very similar to one another however as they grow larger differences in structure and color will become easily apparent. Sometimes the same chemical element can produce different crystals due to the different crystals that the chemical is placed under, an example of this is that diamonds and graphite (the black bit that is in pencils) are both made from an atom called carbon.

What Are Crystals Used For?

Crystals are used for variety of different purposes, some crystals are used for their looks, crystals are often found in jewelry. The crystals are cut down and are called gemstones. They are aesthetically appealing due to their near perfect symmetry as well as their color. However crystals are also used for a large number of other purposes in the modern day.

Watches and crystals have always gone hand in hand. The glass screen that allows you to see the face of the watch is not actually glass it is a crystal. It has be shaped to perfection and highly polished to produce a transparent surface that is strong and scratch resistant. They also help the watch run to time due to the way that the crystals vibrations interact with the mechanics of the watch.

Another use of crystals is in microprocessors, which is the 'brain' of all of your electronic devices. There is a chemical element called silicon, this is the chemical that is greatly used in the production of microprocessor and a number of other electronic devices (thus the name 'silicon valley'). Silicon is produced by melting crystals and molding the liquid into the correct shape so that they can be used in production.

LCD displays are extremely common in the modern world, from computer screens to phones, these are more than likely to be an LCD display. LCD stands for Liquid Crystal Display, which means that each pixel on your screen contains a small amount of a liquid crystal, when electricity is sent through this liquid the chemical structure of the liquid changes. This leads to us being able to control the color of the light that is produced after it has passed through the liquid crystal. This is very convenient however LCD displays must be properly recycled when the appliance has exceeded its lifetime.

Optic fiber is becoming more and more common as a medium for transmitting data. Communication lines made from optic fiber offer the best transmission speeds we have seen to date, these lines are made primarily from glass however glass is not just made from heated sand, it is also made using specific crystals that will allow the fiber to have greater strength.

Finally a large number of crystals are used for spiritual purposes which each crystal having a specific property. There is little or no scientific evidence supporting this beyond the placebo effect however, due to the large quantity of anecdotal evidence this concept is still believed and followed by a large number of people.

CRYSTAL PROJECTS

Salt Crystal Coral

This is probably the most common crystal growing project and will produce crystals with a coral like structure with a white color, however food coloring can easily be added to produce a vibrantly colored crystal structure. Although only household chemicals are used some of them are still toxic and therefore this project should be carried out with adult supervision.

You will need:

- 6 tablespoons of salt
- 4 tablespoons of liquid bluing
- 4 tablespoons on water
- 4 tablespoons of ammonia
- A bowl
- A piece of charcoal or broken clay flowerpot
- Vaseline
- Food coloring

The Method:

Mix together 4 tablespoons of salt, liquid bluing, water and ammonia together. Next take your bowl and coat the inside of it with Vaseline. This will help the crystals to grow on the charcoal or clay. Then place your charcoal or clay in the bowl, then you will need to pour the mixture over your charcoal or clay you may choose to place drops of food coloring on to the clay or charcoal after the liquid has been poured on in order to change its color. Leave the bowl for a day and then add 2 further tablespoons of salt to the clay or charcoal.

Crystals will begin to grow within 6 hours and will continue to grow for three days. After these three days you can repeat the process of adding more of the mixture to the already grown crystals to make them continue to grow.

The science:

The clay or charcoal is porous, meaning that it has a large number of tiny holes that suck up the liquid. The water will evaporate from the surface of the porous material and the chemicals are left behind forming crystals. However these crystals are extremely fragile so make sure to not move the bowl until you are done with the project.

As evaporation is the key factor in this project you should place the bowl somewhere with low humidity and good air circulation. You should also ensure the bowl is not in direct sunlight as this will cause the liquid to evaporate too quickly and not produce the crystals you are looking for. The growing liquid in this project I have found to be very easy to clean, the liquid will not stain your bowl from my experience.

Epsom Salt Frost Crystals

This is widely regarded to be the easiest and fastest way to grow a crystalline structure, in just a few hours you can have a colorful beaker of spikey crystals. Although Epsom salt is not toxic it can have unwanted affects if consumed, also we will be using hot water therefore this project should be carried out with adult supervision.

You will need:

½ a cup of Epsom salt

½ a cup of very hot tap water

A beaker

Food coloring

The method:

Take you beaker and fill it with your Epsom salt and hot water, then leave the mixture for a few minutes. This will ensure the salt has absorbed into the water. Then you can add your food coloring in order to color your crystals. You will then need to place your mixture into the refrigerator. Wait a few hours and you will see that the beaker is now full of crystals. Carefully tip of the remaining liquid from the beaker and you can now examine your fully formed crystals.

The science:

Epsom salt is actually a chemical called magnesium sulfate, when the water is heated it can dissolve more of the magnesium sulfate this will create a 'saturated solution' meaning that no more salt can be absorbed by the water. When you place mixture in the fridge it cools the liquid very quickly which encourages crystal growth. This is as when the mixture cools it becomes denser, and as it becomes denser the Epsom salt particles bond with each other creating a crystal structure.

Growing Alum Gemstones

This project will produce single large white crystal that have the appearance of a natural gemstone. We will be using hot water therefore this project should be carried out with adult supervision.

You will need:

Alum (which can be found in a stores spice aisle)

A beaker

A shallow dish

A pencil

Fishing line

In a beaker pour ¼ of a cup of very hot tap water, then you will need to slowly add and stir alum into the water until no more of the alum will dissolve in it, this is your saturated solution. Then pour the mixture into a shallow dish and clean out your beaker. When pouting your solution into the shallow dish be sure that you do not get any solid alum in the dish, only the solution. Leave the bowl undisturbed overnight and the next day you will see crystals have formed. Carefully pour off the liquid and choose the best crystal, take this crystal aside.

Take your clean beaker and make another water alum saturated solution ensuring there is no undissolved alum at the bottom of the beaker. Tie the fishing line around the crystal you set a side earlier and tie the other end of the line to the pencil. You want to tie the string to the pencil at a length that when suspended in the beaker the crystal will not touch a sides of the beaker.

Suspend the pencil across the top of the beaker with the solution in it. Cover the jar with a paper towel and leave the crystal to form, once the crystal is at a size that you are happy with you can carefully pour away the remaining solution and untie the crystal dry the crystal and you have your final alum crystal.

The science:

The smaller original crystal that was formed in the shallow dish is called the seed crystal and formed because of a process called nucleation. This is the process of a few molecules forming together in a crystal pattern, this structure continued to grow to form the crystalline solid that you can see. They would continue to grow until all of the alum in the solution was absorbed.

Rock Candy Sugar Crystals

These sugar crystals are also called rock candy and are actually a hard edible crystal. This is as the rock candy is made primarily from sugar, you may choose to add color and flavor to your sugar crystals. We will be using hot water therefore this project should be carried out with adult supervision.

You will need:

Three cups of sugar

A cup of boiling water

A beaker

A wooden skewer

Flavoring

Food coloring

The method:

Take you cup of boiling water and dissolve all of the sugar into the water, you will want to keep the water boiling while mixing the sugar into the water so you should use a stove to keep the waters temperature up. Once the sugar is fully dissolved you will need to add and mix in your flavoring and food coloring to your mixture. Allow the liquid to cool somewhat then pour it into its container, remember the liquid will still be hot at this stage so ensure that you are careful.

Now take your wooden skewer tie it so something that can be placed across the top of the jar to ensure that the skewer does not touch the sides of the jar. Place the jar somewhere where it will remain unmoved as not to disturb the crystals forming. You will also need to use a paper towel to cover the top of the jar.

The crystals may take a few days to grow, if you see crystals growing at the top of the jar you should remove them and you can eat them. This is as they will compete to grow with the crystals growing on your skewer. Once you are happy with the progress of your rock candy you can remove it from the jar and eat it.

The science:

When the water is heated it can dissolve more of the sugar this will create a 'saturated solution' meaning that no more sugar can be absorbed by the water. When you place mixture in a place where it will be undisturbed it will encourage crystal growth. As times goes on the sugar particles will find each other on the skewer and gradually it will becomes denser. The sugar particles bond with each other creating a crystal structure.

Table Salt Crystals

These are one of the easiest crystals to grow and produce. You can change the appearance of this type of crystals and this will be discussed in the method. We will be using hot water therefore this project should be carried out with adult supervision.

You will need:

Table salt

Water

A beaker

String

Pencil

The method:

Take your cup of boiling water and dissolve all of the salt into the water, you will want to keep the water boiling while mixing the salt into the water so you should use a stove to keep the waters temperature up. Once the salt is fully dissolved you will need to add and mix in your food coloring to your mixture, if you would like colored crystals.

You want to produce your crystals quickly you should

take a piece of cardboard and soak it in your mixture, then leave the cardboard out in the sun and small crystals will appear on the surface of the cardboard within a number of hours.

If you would like to see a large number of crystals you should leave your beaker over a longer period of time and let the liquid gradually evaporate. This will give you a large number of crystals as the crystals have a large amount of time to form during the evaporation process.

To grow a cuboid crystal you should move the liquid from a beaker into a shallow dish and let it evaporate, this will produce a cuboid crystal.

    If you would like to grow one larger crystal you should take the step for making a cuboid crystal as described above and then tie the string around the crystal you have selected and tie the other end of the line to the pencil. You want to tie the string to the pencil at a length that when suspended in the beaker the crystal will not touch a sides of the beaker.

Suspend the pencil across the top of the beaker with the solution in it. Cover the jar with a paper towel and leave the crystal to form, once the crystal is at a size that you are happy with you can carefully pour away the remaining solution and untie the crystal dry the crystal and you have your final salt crystal.

Blue Copper Sulfate Crystals

These crystals are naturally blue and hold a vivid color. Although only household chemicals are used some of them are still toxic and therefore this project should be carried out with adult supervision.

You will need:

Copper sulfate

Water

Beaker

The method:

The easiest method to create copper sulfate crystals is to add your copper sulfate to your water in a beaker, let it evaporate and you should have some crystals. This being said, I do not believe this is the best method. What I would recommend for you to do it detailed in the following paragraphs along with a few safety considerations that should be had due to the toxicity of copper sulfate

Firstly you need to take your boiling water and dissolve all of the copper sulfate into the water, you will want to keep the water boiling while mixing the copper sulfate into the water so you should use a stove to keep the waters temperature up.

Then you should move the solution from a beaker into a shallow dish and let it evaporate, this will produce a small 'seed' crystal. You will need to take your seed crystal and place it into a beaker with the remaining saturated solution. Check on the solution twice a day and remove any other crystal growths that aren't attached you your seed crystal. This will ensure your seed crystal grows to its maximum size. Once you are happy with your crystal size you can remove them from the solution and allow it to dry.

If you wish to keep your crystals you should keep the crystals in an air tight container as the crystals contain water and will evaporate if stored incorrectly. If the water within the crystals evaporates then the crystals will lose color.

Some safety information that should be taken into account when carrying out this project follows. Do not drink or consume anything with copper sulfate in as it is toxic to you. Wear goggles and gloves to ensure none of the solution or raw chemical touches your skin. If you do get it on you please ensure you immediately wash your hands thoroughly.

Blue-Green Copper Acetate Crystals

There is an easy method of creating crystals using copper acetate, the structure of these crystals will look like an uncut gemstone and has an interesting blue-green color. Although only household chemicals are used some of them are still toxic and therefore this project should be carried out with adult supervision.

You will need:

Copper acetate monohydrate

Hot water

And possibly vinegar

The method:

First of all you need to take 200ml of boiling water and dissolve 20 grams of copper acetate into the water, you will want to keep the water boiling while mixing the copper acetate into the water so you should use a stove to keep the waters temperature up. You will have a saturated solution when the water does not take on any more of the copper acetate.

If there appears to be undissolved material inside for your beaker, you should take a few drops of the vinegar and stir them into the solution. This should get rid of any of the unwanted material.

You should then place the beaker in an undisturbed location to allow the crystal to begin its growth. This process can take longer than some of the other examples and you may begin to see crystal growth after a few days of leaving the mixture. Once all of the liquid has evaporated you should have a large crystal structure in the base of your beaker.

If you would like to grow one large crystal then you should move the solution from a beaker into a shallow dish and let it evaporate, this will produce a small 'seed' crystal. You will need to take your seed crystal and place it into a beaker with the remaining saturated solution. Check on the solution twice a day and remove any other crystal growths that aren't attached you your seed crystal. This will ensure your seed crystal grows to its maximum size. Once you are happy with your crystal size you can remove it from the solution and allow it to dry.

The science:

    The smaller original crystal that was formed in the shallow dish is called the seed crystal and formed because of a process called nucleation. This is the process of a few molecules forming together in a crystal pattern, this structure continued to grow to form the crystalline solid that you can see.

Big Borax Colored Crystals

You can grow these crystals in any color and they will grow to create amazing chunks of vivid color. These are also very easy crystals to make. Although only household chemicals are used some of them are still toxic and therefore this project should be carried out with adult supervision.

You will need:

Borax

Water

A beaker

Colored pipe cleaners

Colored food coloring

The method:

First of all you need to take boiling water and dissolve as much borax into the water as you can, you will want to keep the water boiling while mixing the borax into the water so you should use a stove to keep the waters temperature up. You will have a saturated solution when the water does not take on any more of the borax.

If there appears to be undissolved material inside for

your beaker, you should transfer the solution to another container to remove the unwanted material.

You should also add a 10 drops of food coloring into your mixture so that when the crystal forms it is of the desired color.

You then want to take you pipe cleaner, which should be the same color as your food coloring, and bend it into which ever shape you would like the crystal to take. It should be the same color as your food coloring so that the pipe cleaner cannot be seen after the borax crystal is fully formed. Tie the other end of the pipe cleaner to the pencil. You want to tie the pipe cleaner to the pencil at a length that when suspended in the beaker so the crystal will not touch a sides of the beaker.

Suspend the pencil across the top of the beaker with the solution in it. Cover the jar with a paper towel and leave the crystal to form, once the crystal is at a size that you are happy with you can carefully pour away the remaining solution and untie the crystal dry the crystal and you have your final borax crystal.

One thing I could recommend if you are going to carry out this method is that the shape that you bend the pipe cleaner into, my favorite shape is to bend the pipe cleaner into a disk shape as I find that this produces a large crystal with interesting crystal growths on its surface.

Aragonite Crystals

These crystals are very unique from the other crystals that I've talked about as they do not involve the normal saturated solution. Instead the base of theses crystals is dolomite rocks.

An adult will most likely need to order these rocks for you and therefore this project should be carried out with adult supervision.

You will need:

    Dolomite rocks

    Distilled white vinegar

    Beaker

The method:

You will often have to buy your dolomite rocks in a larger quantity so maybe find a few friends who would also like to try out this project so that you can split the cost of the rocks and ensure they are all used. Dolomite rocks are a sedimentary rock, this means that it's made from sediment and minerals from the bottom of the ocean.

Choose one of your rocks to carry out the task and place it in your beaker. Then take your vinegar and fill up the beaker so that some of the rock is still exposed at the top of the beaker.

You should then leave the jar in an undisturbed location for around two weeks or as long as it takes for the vinegar to evaporate from the container.

If you disturb the container before this point you may break the crystals that are starting to form on the surface of the dolomite rock. However if you see crystal growth starting around the edge of the container it would be wise to remove it as this will detract growth efforts from your main crystal.

You will start to see some good levels of crystal growth after around 5 days. To speed up the evaporation it will not hurt to place the jar in a sunny spot such as a window ledge.

Once the crystals have formed and all the vinegar has evaporated you will have your final fully grown crystal which you can now keep and store.

Color Changing Crystals

This project will produce single large crystal that have the appearance of a natural gemstone and will change color from green to yellow depending on the light and temperature. Although only household chemicals are used some of them are still toxic and therefore this project should be carried out with adult supervision.

You will need:

10 grams of Alum (which can be found in a stores spice aisle)

A beaker

A shallow dish

3 grams of red prussiate

50 milliliters of water

In a beaker pour 50 milliliters of boiling water, then you will need to slowly add and stir alum and red prussiate into the water until no more of the chemicals will dissolve in it, this is your saturated solution.

Then pour the mixture into a shallow dish and clean out your beaker. When pouting your solution into the shallow dish be sure that you do not get any solid alum in the dish, only the solution. Leave the bowl undisturbed overnight and the next day you will see crystals have formed.

Once you are happy with your crystal size you can remove it from the solution and allow it to dry. You should then dispose of the chemical solution.

The best way to see the color changes in your crystals is to take the crystals and place them into two containers, one of the containers must be transparent. You should then place the transparent container with the crystals in it into a location with a healthy amount of sunlight, such as a windowsill. Place the other container in the dark and over the following days you should check on the color of each bath of crystals.

What you will find is that the crystals that were left in the dark will remain to be a yellow color however the crystals that were left in the sunlight will change color from yellow to green and then finally to a blue color. This reaction may take place over the course of a few days or could happen within a few hours.

The chemicals that are required if you are going to carry out this project are safe to use however in excess some of the materials could be toxic, ensure to correctly store and handle your chemicals during the project.

Bismuth Metal Crystals

A bismuth crystal is a metal crystal and it is also the easiest and fastest metal crystal to grow at home. The crystals have a strong geometric patterned visually are quite stunning, they have a rainbow like appearance on their surface and are shaped almost like pyramids. Although only household chemicals are used some of them are still toxic and therefore this project should be carried out with adult supervision.

You will need:

Bismuth

Two aluminum cans

A stove

The method:

Firstly you will need to obtain your bismuth this can be found easily online so be sure to order the bismuth well in advanced of you wanting to carry out this project. You will also need to take your aluminum cans and cut them in half's so that they form cups or bowls.

The pretense to growing bismuth crystals is that it has a fairly low melting point, which sits around 520°F. This makes it easy to place bismuth into your aluminum cup and melt the metal over a stove, then once the metal is melted it will begin to crystalize and separate its self from its impurities. At this point you will need to separate the impurities from the crystal and this requires some good timing, however this will come with practice. If you do not get this right you can always re-melt the bismuth and try again.

The first step in this process is to place your bismuth into one of your container and place it over a high heat on a stove until the metal melts completely. You will need to wear gloves and goggles as you do not want to be working with molten metal without them. A grey coat will form on the top of the molten bismuth.

You will then need to heat up another container and carefully pour the molten metal into the new container, you want to pour the bismuth gently and only pour it so that only the material from under the grey skin is transferred.

Place the container with the newly transferred pure bismuth onto a heat resistant surface and then wait for the crystals to form through the gradual cooling of the container and the molten metal itself.

After around half of a minute you will see that the bismuth has begun to crystalize. At this point you will want to pour the liquid bismuth that is still cooling on the top of the crystals away and into another container. You will know when to do this the stage of cooling when the bismuth has crystalized but will still shake when moved.

Once the crystals have finished cooling down the metal will be very brittle and so you can easily snap the crystals away from the can container. If you are unhappy with the way your crystals came out or fancy having another go at this project you can simply re-melt the bismuth and try again with a new set of can containers.

Sodium Nitrate Crystals

This is a chemical that can be found commonly in food, fertilizer and also in explosives, this forms very interesting crystals which are colorless without the addition of a food coloring. This project will form a hexagonal structure of crystals and are a bit more complex than some of the other crystals that I have described. Although only household chemicals are used some of them are still toxic and therefore this project should be carried out with adult supervision.

You will need:

   110 grams of sodium nitrate per 100ml of water

   A beaker

   String

   Shallow dish

The method:

First boil your water and add your sodium nitrate, the ratio of sodium nitrate to water that is listed above however you may need to tweak it depending on your situation, keep adding it until no more will be absorbed. This will mean you have created a saturated solution. You will then need to add your food coloring if you are choosing to add some into your crystals.

One way to proceed from here is to then simply leave the water to evaporate and it will leave you with crystals.

Another method of getting the crystals to grow is to prepare the saturated solution as instructed earlier, you then need to allow the solution to cool completely. Then you must add a further few grains of sodium nitrate to your solution and then seal your container. Leave the container for two days and this will ensure that the solution is fully saturated.

Pour the mixture into a shallow dish and clean out your beaker. Leave the bowl undisturbed overnight and the next day you will see crystals have formed. Carefully pour off the liquid and choose the best crystal, take this crystal aside.

Take your clean beaker and make another water saturated solution ensuring there is no undissolved sodium nitrate at the bottom of the beaker. Tie the fishing line around the crystal you set a side earlier and tie the other end of the line to the pencil. You want to tie the string to the pencil at a length that when suspended in the beaker the crystal will not touch a sides of the beaker.

Suspend the pencil across the top of the beaker with the solution in it. Cover the jar with a paper towel and leave the crystal to form, once the crystal is at a size that you are happy with you can carefully pour away the remaining solution and untie the crystal dry the crystal and you have your final crystal.

Further Projects

There are a number of factors that can affect the success of your crystals growing, one of the most common problems is the environment that the container is left tin while the crystals are forming. You could do an experiment in order to see what conditions are best for crystal growth. The two main factors that should be taken into account in an n experiment like this is the 'temperature' and the 'humidity' of the growing environment.

Another thing you could test for is what effect the rate of cooling has on each of the projects crystal structures.

FINAL THOUGHTS

Thank you for purchasing this book. Hopefully this book has fulfilled the role of aiding you in your crystal growing ventures. I have always been fascinated with the idea of being able to grow a crystal. And I am glad to share my knowledges and experiences with you so that you too may grow many wonderful crystals. I would love to hear some feedback from you to see what you thought of my efforts and project ideas.

Made in the USA
Monee, IL
23 April 2021